International Farm Stock Library

OLD
&
RARE BREEDS OF CATTLE

OTHER TITLES

The publishers specialize in rural books with special reference to poultry breeds and poultry management. Under the aegis of The World Bantam & Poultry Society advise is given and books published which deal with the return to Free Range and allied methods, other than battery cages.

BREEDS OF CATTLE

Old & Rare

J. BARNES

Beech Publishing House
7, Station Yard
Elsted Marsh
MIDHURST
West Sussex GU29 0JT

© BPH Ltd, 2003, 2006

This book is copyright and may not be reproduced or copied in any way without the express permission of the publishers in writing.

ISBN 1-85736-134-2

This edition 2003
New Impression 2006

Part previously published as
**Breeds of Cattle
and How to Know Them**
E C Ash

British Library Cataloguing-in-Publication Data
A catalogue record for this book is available from the British Library.

Beech Publishing House
Station Yard
Elsted Marsh
MIDHURST
West Sussex GU29 0JT

CONTENTS

Chapter 1 Breeds Explained   7
Chapter 2 The Nature of the Cow   69
    Index   80

FOREWORD

Cattle are an essential part of our countryside and have been bred for hundreds of years. Sadly many of the older breeds have become quite rare and some appear to have been lost for ever.

This book sets out to give details of the different breeds in the hope that it will give encouragement to try to bring back those which are in danger of extinction. At the same time, it should be appreciated that some of these animals can be profitable, even on the most difficult tracts of land.

Chaper 1

The Breeds Explained

NOTE:
Milk yields and weights are given as a guide, but may be different for individual breeds and strains kept today.

ABERDEEN ANGUS
(Hornless: a beef breed),
Colour : Black

This beautiful creature, so full of quality that its black coat sparkles, as if it has been polished, is the outcome of many a cross of early breeds, and although hornless, built up on a horned cow, very similar to the Kerry in appearance.

In the early history of the breed a short legged, hornless animal from Scandinavia was introduced and mated with this Kerry-like cow; and, later, many other breeds all played a part, resulting in the Aberdeen Angus of to-day, one of the most magnificent things on earth.

One of the first and most important of the breeders who. made *Doddie* history (for Aberdeens are known as ' Doddies') was Hugh Watson, who in 1808 developed the Aberdeen by skilful breeding, Later M'Combie and then Sir George Macpherson Grant did great work in the same direction, and it is these three breeders who deserve the fame of having produced the Aberdeen Angus.

From so lowly a beginning, and a horned cow too, the Aberdeen Angus cattle rose to be one of the finest beef breeds existing, maturing early and producing flesh of excellent quality. Of a dark black colour, with a deep and broad body and short legs, their quality resembling that of the Red Poll, their rapid fattening powers and hardy constitution made them peculiarly suitable for the foreign meat trade.

BREEDS OF CATTLE

Polled Aberdeen-Angus Bull *(Justice)*

Being hornless means that they can be en-trained much more easily than the horned breeds, and can travel long distances without injuring each other.

In breeding these cattle, it is usual to allow the cow to rear her own calves, so that the calves run with their mothers; and as the calves grow older, cake and corn are added, for early beef is aimed at, and the quicker the animal matures the better, of course.

Although a North-country breed, many famous herds are kept in England. It is interesting that when the Aberdeen is used for crossing purposes, even with the oldest horned herds of cattle, the result is a hornless black or a blue roan. The cross adds early maturing powers to the progeny, and hence these crosses are in great demand, and a blue roan bullock usually makes more than any other colour.

The approximate weight of Aberdeens was as follows :

Steers under 2 years 1,400 lb.

Steers under 3 years 1,835 lb.

Steers over 3 years 2,170 lb.

Smithfield Club Show (4 years' average), but modern figures would show different weights.

Note: Not now as popular and have been modified so some strains are taller and longer.

BREEDS OF CATTLE

Aberdeen Angus Cow *Juana Briga*

Aberdeen Angus Bull

AYRSHIRE
(Horned: a milk breed).

Colour: White marked with clear browns.

The Ayrshire is not an ancient breed, and it was developed from Dutch cattle, Shorthorn, and possibly Highland cattle all intermixed. It is named after the county of Ayr, and is one of the finest dairy breeds existing. During its history it was known as Dunlop cattle, and also as the Cunningham breed, after the area in which it developed. In 1877 the Ayrshire Cattle Society was formed, and from that date onwards the popularity of this variety increased and a good export trade followed.

The milk is different to other milk in that it has fat in very small globules, which makes it particularly suitable for cheese-making, but hardly suitable for butter-making.

Ayrshires are not large animals, the weight of a cow being about 1,000 lb.; but they are heavy milkers, and the record is, I think, a 2,000-gallon cow, while 1,000 to 1,500 gallon cows are very numerous.

The colour of the animals is white, marked with distinct blotches of brown and grey red. In shape they are typical dairy animals, with wonderful udders. The lacteals, which were once upon a time the subject of a fad, and were made so small (especially the hinder two) that they were really too small to milk, have now been enlarged, and are the usual size. The horns are long, turned up (see illustrations) and the nose is black or pink.

When the horns did not appear to be growing at a correct angle, they were fastened to pulleys on which a weight is attached; at first 1.50 lb. is sufficient, and this was gradually in-

BREEDS OF CATTLE 13

creased to as much as 3 lb.

Ayr, Dumfries, Castle Douglas, Paisley, Stranraer, were centres for the Ayrshire breed, but they are also to be seen at the Royal and Dairy Shows.

The bull calves are of no use to keep for beef, and the cow, when no longer able to milk, has also no beef value; serious as this may appear in print, in actual fact the huge milk yields make up for this failing in beef qualities.

Ayshire Cow (c. 1860)

Ayshire Bull from an old print about 1860
Many bulls portrayed in the 1840s were much deeper and heavier.

Blue Albion

BLUE ALBIONS (*Milk and beef breed*).

Colour: White with blue markings, blue roan.

The first exhibition of the Blue Albions at the Royal Show caused considerable interest, for these cattle were a new breed as far as the ordinary live stock pedigree world was concerned.

It is said that they had lived on the hills of Derbyshire, especially in the Peak district, for a large number of years, and were extremely hardy, able to stand the hard conditions of life on those hills.

They probably originated from a cross of Roan Shorthorns with the Friesian, for this cross gives similar cattle.

They have the good points of both these breeds, – the constitution of the Shorthorn and the wonderful udders of the Friesian.

Their milking powers was said to be good, and as much as 12,245 lb. of milk was a record of a cow living on the hills. If they can milk this on the rough feeding of the Peak area what would they be able to accomplish under proper milk feeding as carried out in the great milk-producing areas of this country?

The colour of these animals is a blue roan with a blue or black nose. They are the size and have the shape of the dairy Shorthorn, and carry exceedingly well-shaped udders.

The Blue Albion Cattle Society was formed in 1920.

THE NORTH DEVON
(Horned: milk and beef breed).
Colour: *Brown* red.

Wonderful square cattle and a type of their own, the North Devonshire curly-coated red breed, with comparatively long horns, are believed to have been brought here by the Phoenicians when these people visited our shores to obtain the Cornish tin. It was certainly very kind of them to bring such grand cattle with them.

Devons are hard cattle, able to thrive on a low ration; in fact on many farms they are not caked either summer or winter, and they must be delighted when a couple of lbs. of cotton cake are given them when the grass is young. They only get it when they show they need it, in order to counteract the liquid condition of this feeding.

Their milk is of such rich quality that the breeders claim that 2.1 gallons of milk will make 1 lb. of butter. Living as they do out in the open air, the cattle are exceedingly healthy, and cattle tuberculosis is rarely met with.

Their hardiness and thrift have resulted in a good demand from South Africa as well as the rest of the world, and it is said that this breed is peculiarly able to stand the droughts met in some overseas countries.

For crossed bred cattle, Devon, crossed with Angus, are very popular; the high quality of the Angus and the hard thrifty constitution of the Devon results in an ideal breed for grazing and fattening.

AYSHIRES & POLLED ABERDEEN

Dairy Shorthorns

BREEDS OF CATTLE

A problem with the breed has always been the poor milk yield, and they also suffer from a slow rate of fattening. There was a feeling long ago that they would have to be amalgamated with shorthorns.*

Old Style North Devon Bull

---

* *British Farming*, John Wilson, Adam & Charles Black, Edinburgh, 1862.

DEXTER

(Horned : a milk and small beef breed)
Colours: Reds and Blacks.

The Dexter is an Irish breed, a very small animal indeed – the smallest thing you have seen in cattle. Its size makes it most fascinating, and at the shows visitors are most anxious to see the Lilliputians of the cattle world; and the crowd you see is often on its way to the Dexter sheds to look and admire and wonder where such cattle could come from.

It has the tiniest of limbs; the length of foot to knee is about eleven inches, and the whole front leg is only 23 inches long ! and from the ground to its body is a mere 21 inches.

Their colours are always black or red, and they have horns rather like shorthorns.

Except that it is known to be a very old breed, its history is lost, and it seems strange that so small a variety, a veritable bantam of the cattle world, has been able to hold its own when size in animals has been so much in demand.

From a commercial point of view, the value of the Dexter lies mainly in its heavy milk yields from so diminutive a body. Although such tiny things, some Dexters give many gallons of milk a day-a wonderful accomplishment for a cow six feet long, weighing 8 cwt. Of course for milking, a very low stool is needed. Unfortunately monstrosities, bulldog calves, the most weird things, are often born, but to-day the number of these bulldog calves is less than it was. These strange calves resemble sun-fish more than cattle, and occasionally a bulldog calf will be produced, and then some weeks later the same cow

BREEDS OF CATTLE

Above: Dexter Bull *Pontarfran Rifleman*
Below: Dexter Cow *Pontarfran Patti*
These are fully grown "bantams" of the cattle world (8 cwt)

will give birth to a proper calf, healthy and fit.

Dexters are certainly very sweet things, and have wonderful udders for so small an animal. When crossed with larger breeds they make small quick-growing stores, ie; bullocks to be fattened for beef.

THE GLOUCESTER BREED
(Horned: milk and beef cattle).
Also known as OLD GLOUCESTERSHIRE
Colour : Red brown with white and black.

One of the oldest pure breeds of the world, the old Gloucester cattle of a rich dark brown colour are one of the most interesting. These cattle are marked with white on the brisket, udder, and a white line along the back and a white tail, whilst the head and legs are black. They are very beautiful.

The following are the chief characteristics:
Body of a black brown colour, head and legs black, white streak on back, white tail, long hair and bushy, white belly, black teats, and the top of the tongue black.

The great age of this breed is proved by the fact that once one of these cattle has been used to cross with any other breed, the white line appears in the progeny not for that generation only, but for many generations without any further infusion of this breed.

In 1834 the breed was nearly extinct. and it was thanks mainly if not entirely to the Duke of Beaufort that this variety continued. In 1919 a society was formed to revive the breed, but into the 1950s there was still little progress.

Note: Unfortunately the breed is now very scarce and efforts are being made at revival.

BREEDS OF CATTLE

Gloucesters
Above: Bull *Badminton Bentley Star II*
Below: Cow

FRIESIANS
(Horned : a milk and beef breed) .
Colour :-Black and white.

This is one of the leading dairy breeds of the world, with a history which is astonishing, for at the time, the Friesians, then known as the Dutch, were so unpopular that they made very little money. Farmers were so deeply prejudiced against this variety -- they had the reputation of giving poor quality milk, which sooner or later led to police-court prosecutions, that they would not buy them. But in 1909 the Friesian Cattle Society was formed, and from that date the history of the breed was one of constant development, not only in the prices obtained at auctions, but also in the quantity of milk yielded and the quantity of butter fat produced, and the size and shape of the cows. There were at one point 54 x 2,000-gallon cows of this breed in Great Britain, and one reached 3,000 gallons. As for butter fat, 1,316 lb. of butter has been produced by a single Friesian in a year, and they have won honours for fat percentage at the Dairy Show.

The importation of Dutch cattle from Holland under special licence was a move which has had the most remarkable results, making the Friesians huge, grand cattle, beautifully marked. A further importation of 1922 of South African Friesians caused a stir in cattle-breeding circles, for never have such prices been realized for cattle as were realized that day.

At the Dutch imported sale, 59 head averaged £253,. the highest price being 560 guineas for *Robert* , and the same price for *Lyn*, and 520 guineas for the Cow *Betje IV* ; and high as these prices were the sale in 1922 of South African Friesians beat all expectations, for the average was £1242 odd for 83 head, the highest price being £3,900 for *Harlens Marthus,* a yearling and bull , 4,300 for *Melrose Diliana,* a two-year-old cow.

The colour of the Friesian is black and white ; mouse coloured and entirely black Friesians do occur, but are not so popular. The black should appear in bold patches. The Friesians are large, well-shaped dairy animals, with short horns and black noses, and it is remarkable how docile the bulls are, just like children.

British Friesian Cow.   *Kingswood Myrtle*
This cow gave 2281 gallons in 365 days

BREEDS OF CATTLE

British Friesian Bull
Kingswood (imp) Ynte
Imported from Holland by Mr Horace Hale. Daughters of this bull averaged over 1,000 gallons with their first calf.

Meat Production

The Friesan is probably the best of older dairy breeds for meat production. But it does require good quality pasture or the milk and meat yield will suffer. When the land is poor and at high elevations the Northern breeds such as the Ayrshire have been found to be better.

Note: This continues to be a popular breed. However, the Holstein has been produced which is similar to the Friesian.

GALLOWAY CATTLE
(Hornless.. a beef breed).
Colours: Black, red and belted.

This very hardy breed is one of the oldest pure breeds existing, and its earliest history, like that of all cattle, is unknown, but it is very probable that they are closely related to, or may actually be, the Norse cattle brought to England by the Vikings and other visitors of those early times.

We read that in 1530 Galloway bullocks were sold at 8d. each, 8d. being the price of a yearling, whilst two-year-old bullocks made 11d., and cows 22d. (It must be remembered that a penny then was worth far more than it is to-day). It was customary during the 18th century to drive these cattle from the North through England, to be sold as stores; and it was in this way that many of our breeds obtained a touch of Galloway blood, which still shows in their character. Hence the Galloway type of some of the Red Polls to be found in Norfolk.

Galloways are black or red brown cattle, hornless, with black nose. The head has not the pointed top of the Aberdeen Angus and Red Poll, but is broad and flat.

They are a typical beef breed: a hard, well-paying beef-making proposition, especially where fodder is poor or scarce. The cows, although not kept for dairy purposes, rear their own calves.

Besides the black and red Galloways there are two other varieties, one of which is the Belted Galloways, one of the most charming and original creatures living.

Whilst occasionally the marking may not come quite regular, white calves are never born, and only rarely do we find

BREEDS OF CATTLE

an entirely black calf.

Of all the Galloways, the Belted are the best milkers, and their milk is very rich. There is a good demand for belted cattle for Parks, and their marking adds to their value, apart from their feeding qualities. *However, they are now scarce and need to be revived more.*

Lastly, there are also White-faced Galloways, known as *Brocket-faced*, which are found in certain areas.

Weight of steer under 2 years ...1,160 lbs. over 2, 1,665 lbs. over 3 1,990 lbs.

Belted Galloways Grazing. Belted are the best milkers.

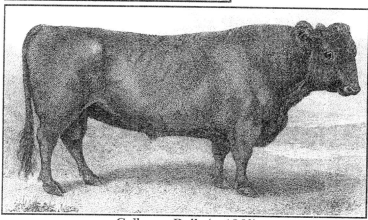

Galloway Bull (c. 1860)

GUERNSEY

(Horned: a milk cream breed).

Colour :-Fawn and brown yellow, marked with white.

Of all bleeds for dairy qualities the Guernsey is one of the most famous, for while being exceptional milkers, and often giving from 1,000 to nearly 2,000 gallons of milk a year, the milk is very rich in fat. If fed on scientific lines the milking powers may be so developed as even to compete with the Friesian as heavy yielders.

On the green fields of Guernsey, on an area of only 12,600 acres, 6,000 head of cattle were kept, and it is customary there to tether the cattle to stakes, moving these stakes as the animal consumes the vegetation around it. Guernseys have a charming appearance ; they are fawn and brown, and the skin is orange yellow, very similar to the colour of the butter produced from their milk.

It is interesting from an historical point of view that, although the island of Jersey is close by, the Jersey and Guernsey cattle have not been interbred. The chance of this occurring was forseen, and great care taken to prevent it. In 1800 a law was passed making importation of breeding stock illegal, and it is thanks to this measure that the purity of the breed has been maintained.

Guernseys are not more popular because the male animals, although they will fatten, have yellow flesh, which is not liked in England, although acceptable in many overseas counties. Many of the steers are not reared because of their limited value.

BREEDS OF CATTLE

Guernsey Cow *(Fi Fi)*

The average weight of a Guernsey cow is about 1,000 lb. They have the appearance of being delicate, but they are not. Owing to its milking powers the breed is very popular in America, and huge records have been made by development on a scientific basis.

Considering the quantity of milk yielded, the great fat percentage and the wonderful colour of the milk and butter, the Guernsey is worthwhile, although not the most popular of breeds.

Guernsey Cow

Guernsey Cow *Ruby First*

HEREFORD (*Horned: a beef breed*).
Colours: *Deep* red and white.

The famous Herefords are named after the county, where during the 17th century, the greatest efforts were made to develop a special breed. Careful selection has resulted in one of the most magnificent beef type of cattle conceivable: in fact, the most famous of all the beef breeds. As a breed they are exceptionally healthy and free from the baneful tuberculosis, and in consequence of its sterling qualities the Hereford has spread all over the civilized world; and wherever they have gone they have proved themselves to be the most perfect of beef breeds, and thus caused an increasing demand.

The result has been constant high prices for the best. A bull was sold in 1922 for £3,750 to the Argentine. With its breadth and length, you can realize that an animal like that, not only can carry the maximum quantity of meat, but will breed stock able to do the same thing.

In the olden days it was the Hereford that was noted as the draught ox, doing all the work on the farm, pulling the ploughs and harrowing the soil, at about half a mile an hour, no doubt greatly to the satisfaction of the laziest workman and to the annoyance of the keen and energetic, who hated to dawdle.

On some estates Herefords were broken to harness, and the onlooker could spend many an enjoyable hour watching the teams harnessed to the yoke, dragging the creaking agricultural implements. After the work was over the bovine team returned to their 'stables' one by one, as docile and as easily controlled as horses! It was an olden-day picture in 20th–cen-

Kerry Cattle

Welsh Cattle

BREEDS OF CATTLE 33

Hereford Cow *(Merriment)* & Calf

Old Style Hereford Bull

tury England.

Herefords are of deep rich red colour, a dark ruby red, while the under parts, the brisket, the tip of the tail, the crest and the face are white.

Hereford Bull *Premier*

Note: The traditional Hereford has lost ground because of the many crosses.

HIGHLAND CATTLE
(Horned: a beef breed).

Colours: Blacks, browns, greys, fawns.

Highland cattle, or *Killoes*, are one of the most picturesque of British cattle, with their shaggy long coats and upright horns, and a wild look as if they really were quite wild, which of course they are not.

This hardy, thrifty, and powerful breed roams on the mountains and on the wildest pastures of Central Scotland, facing bleak conditions, the cold winds, and ekeing out a living under the most strenuous circumstances. They have to rely on their own wits.

They are brown or red or black in colour, or sometimes a grey or fawn. They are a very slow-maturing breed, taking three or four years to grow to maturity and they are a very old breed, too, for they were known as pure in 1700. Old books on agriculture tell of their value, and how they lived among the wild mountains and glens when at times everything was frozen hard, and there was little or nothing to eat.

The milk is rich in butter fat, but they give little quantity, and the calves run with their mothers, and few cows are kept as milkers. As to weight, Smithfield gave the weights as 1445 lb. under three years. old and 1645 lb. over three years old. Living a free life, these cattle hold their own where few other breeds could exist, for they do not need man's attention as most breeds do.

In the days of the past the farmer handed these cattle over to the dealers to take to market, and as the huge herds wandered mile after mile over the mountains and wastes to the

West Highland Cow & Calf
Based on a painting by Gourlay Steell, depicting an old type of Highland Cow. Note the hills in the background and the rough pasture.

lowland markets and the droves would become larger and larger day by day. A farmer would bring his small lot to add to the great herd, then another would arrive; and so it would go on until the huge drove would stretch great distances controlled by men and dogs.

Highland Bull   *Lord Clyde*

JERSEYS

(Horned: a milk (cream) breed).

Colour: Various shades of fawn.

These beautiful cattle, one of the best dairy breeds of the world, are no doubt descended from the cattle of Normandy and Brittany. They have for many years been bred in the Channel Islands, protected, as in the case of Guernseys, by Acts of the State preventing any chance of spoiling so valuable an asset by the introduction of stock of other kinds.

Jerseys are the most charming cattle, with fawn skins and great, large, sensitive eyes, and deer-like heads, and are light and graceful. The skin is yellow, the same colour as their milk, and the butter made from the milk is yellow too, and looks so very nice that Danish imported takes quite a back place ! Unfortunately the bull calves are no use except for breeding purposes, and hence many people do not keep this breed, for very few bulls are wanted.

Considering the size of the animal the milk yields are extraordinary; thousand-gallon cows are quite numerous, and as they are small feeders they produce these high yields of milk at extremely low costs. Apart from this they are butter cows and one of the best breeds for butter-making. The price of pedigree stock can be below their real value, and there is an opportunity for breeders to get the very best at comparatively low prices. It is clearly a breed with a huge future, for the rich milk and the colour of the butter are assets of great value.

It is interesting that the bulls of this breed, in contrast to the Friesian, are so very savage that they are controlled by chains

BREEDS OF CATTLE

Jersey Cow (*Lady Viola*)

fastened over the horns and through the ring of the nose in order to lever in case of emergency. They are so different from their deer-like wives, so gentle and confiding.

Large numbers of Jerseys have been imported to the USA, Australia and New Zealand, and Jersey crosses are in great demand as dairy cattle, for such crosses give exceedingly rich milk and yet remain dual purpose, known as *Whitefaces*. The general scheme of colouring makes them perfectly wonderful things, for the rich red hair is often curly, and their white faces are often marked with dark lines below the eyes, which help them to stand the glare of tropical suns.

Jersey Bull   *Pilgrim*

KERRY
Colour: Black.
(Horned: a milk and small beef breed).

The Kerry is an Irish breed; an old-world breed too, the last of the black-horned cattle of Western Europe, small cattle, thriving on the poor stony lands, excellent Cotter's cows; the very cow for the poor man with a cottage, who has to let his animal find a living as best it can on the bare pasture of the Common or on the hillsides.

You can imagine these cattle on the mountains finding shelter behind the rocks; a hard, semi-starved, yet prosperous race, giving more milk than-one would think on such meagre diet. The lesson is that where they are well fed, great results are obtained from the breed. surprising results indeed for they will then give heavy yields at very low costs. It is said that three cows are able to thrive and yield well and even grow fat where one of the larger breeds would die; so the reason for the popularity of this breed is not far to seek.

Whilst good dairy cattle. Kerrys also make small beef; and Smithfield gave Kerrys under two years old to weigh 860 lb., over two years 980 lb and over three years 1,165 lb. A full-grown cow will weigh about 700 lb to 900 lb and the milk yield varies from 4,500 lb. to as much as 8,000 lb, a year, and several have given up to 10,000 lb. in a year. At the Hattingley herd no cow gave less than 7,000 lbs. and with second calf is kept, and many of their first-calf heifers give 600 gallons and. over in their first year. The milk is very rich. The weight of a Kerry calf at birth is about 40 to 50 lb., but a case is on record of a cow weighing 7 cwt. actually producing a 126 1b. calf.

BREEDS OF CATTLE

Kerry Cattle

Note: This fine breed is quite scarce and needs to be revived to avoid extinction.

LINCOLN REDS
Lincolnshire Red Shorthorn
(Horned : a milk and beef breed)
Colour: Red.

This variety is a red Shorthorn, a good milker of whole red colour and of dairy Shorthorn type.

The old Lincoln cattle on which the present Lincoln Reds are founded were described by Gervase Markham, in a book entitled *A Way to Get Wealth*, published in 1695. He speaks of a fair bull, and mentions that the Lincolnshire cattle were mostly 'pyde' with more white than any other colour; and that their horns were little and crooked, with thighs strong-boned. He also adds that they are indeed fitted to labour and draught.

The present-day Lincoln Red history goes back to the close of the 18th century, when Mr. Thomas Turnell, of Reasby, developed this breed, which were then known as *Turnell Reds*. So gradually a special breed of Shorthorns of distinctive character came into being, red shorthorns of a milking type.

In 1895 the Lincoln Red Shorthorn Society was formed, and the stock was registered and protected. The merits of the dairy breed are that it is a useful farmer's cow, a heavy milker and a good store producer.

Apart from dairy powers the Lincoln Reds are well known for their beef-producing qualities, and steers at less than three years old give 8 to 9 cwt. of meat.

Note: This fine breed is quite scarce because it has been crossed with other breeds.

Lincoln Red Cow   *Petwood Recorder*

Lincoln Red Bull Calf

BREEDS OF CATTLE 45

Lincoln Red Bull

LONGHORNS
(Horned : a milk and beef breed) .
Colour : Piebald.

The Longhorn carries us back to early English days before railways were invented; the days, when coaches creaked along the roads and were often held up by snowdrifts and sometimes by highwaymen, as well as by other nasty incidents common on the road in olden times.

The Longhorn is an old-world breed, with great long horns and curious markings and colours; and it is said, which is quite likely, that the Longhorn goes back to Roman times and contains in its system the blood of many of the extinct old English breeds.

They are a wonderful variety, too, as a commercial proposition, giving four to five gallons of milk a day of the very richest quality, with as much as 5% of fat. Moreover, they are so hardy, thrifty, and good-natured that they are easily looked after. Once very popular, their day passed as the newer breeds gradually pushed further and further forward, and so powerful were the influences against them that at one time it seemed as if the Longhorn would be pushed aside altogether. But the old breed kept going somehow !

The modern type of Longhorn is largely due to Mr. Bakewell, of Dishley, in Leicestershire, so well known as a live stock breeder that in those days all roads in midland England led to Dishley, if you were interested in breeding animals.

When railing cattle started and it became usual to move live stock in this manner the extraordinary long horns of this breed made it so difficult to pack them into cattle trucks, that it

BREEDS OF CATTLE

no doubt caused their downfall. It was a pity that so highly fascinating and original a variety, with such valuable utility points, should have lost ground. Business advantages have a sad way of turning the dice, and it must be so in the long run, however interesting a thing may be to the general public or even the farmer who must consider the economics of a breed.

Consider horns six feet from tip to tip, and you will realise the disadvantage. No wonder that when these cattle go into buildings they turn their heads side-ways to prevent their horns catching against the doors.

When you see these cattle you will feel that some how or other they are out of place amongst the stock and machinery around them, for they are so old, and everything else seems so young in comparison.

Engraving of a Longhorn Bull

BREEDS OF CATTLE

Robert Bakewell (1726 -- 1795)
A large man who started cattle breeding at twenty years of age and became famous for the stock he produced. He was criticized because of his secrecy, but was still a great animal breeder.

● ● ● ● ● ● ● ● ● ● ● ● ● ● ● ● ●

And you will notice that their colour is partly white, with curious pied and brindled marks, the sort of thing which might go with a pantomime at Christmas time, and which seems so unreal in a farm-yard.

But although the Longhorns are a disadvantage the breed yet has, a future, for, accustomed as it is to the hard conditions up North, it does surprisingly well in an easier life, and moreover will give stamina and health to breeds which may need these characters.

BREEDS OF CATTLE

THE RED POLL
(No horns: a milk and beef breed).
Colour: Red.

The story of the Red Poll started over 200 years ago, when the farmers of East Anglia crossed a Suffolk Dun Cow, a wonderful milker, but lacking constitution, with the beef cattle, the red-horn cattle, of Norfolk ; a breed of poor milkers but great beef cattle. From then onwards the breed had been in the making, and many herds reached success, having the very best dual purpose stock, good milkers, and also able to fatten readily whe desired.

In the story of the Red Poll, the work of Garret Taylor, Alfred Smith, Mr. Hill, and Martin Longe have passed down to history, with several others who kept milk development in view, and bred for milk without losing the constitution of the breed.

In the history, certain famous bulls have pressed type, and amongst them are *Iago, Redmond,* and *Majiolini 3,600.*

Red Polls are rich red coloured, medium sized animals, with short legs. They have no horns, and are born and grow up hornless; the nose is white, and as milkers certain lines are very useful and profitable dairy animals, with a milk record of 800 to 1,500 gallons a year. They are typical dual purpose too, making good beef. The quality of milk is above the average, and a fair yield of good quality milk is able to be produced at low costs. It is on account of this that the breed made great headway.

Red Poll calves at birth are simply charming, and in this breed good breeding shows from the first moment to the last.

Red Poll Cow

Red Poll Bull

SHETLAND

Shetland cattle, while very small, are not quite so tiny as the Shetland ponies. They are pretty little animals, well formed, with stout coats. In frame they resemble the Jersey cattle, and, in fact, are supposed to have some relation to them. Possibly they are descended from the Scandinavian cattle, and it is probable that at some time or other an intermixture of Dutch blood took place.

For a long time, however, there has been a pure Shetland breed, and, as in the case of the Channel Islands the fact of their being separated from the mainland has helped to retain this purity.

It is said that they have a considerable resemblance to Kerry cattle, but whether there is a link between the two breeds is not recorded.

It is said that, for its size, the Shetland milks well. However, it was always a rare breed.

Note: In fact, the position is said to be critical and the Rare Breeds Trust has recorded that there are less than 150 breeding females.

Park Cattle

PARK CATTLE
(Horned and Polled)
Colour : White.

This breed is said to be descended from the old Roman white cattle, descendants of which are still to be seen in Tuscany and other parts of Europe.

They have been bred on the Woodbastick Hall Estate for over a century, and locally they are in great demand for beef. Their milking qualities are also good, just under 1,000 gallons of milk being given by one animal with her second calf.

Two Kinds

There are two kinds: the horned and the polled. They are indeed beautiful creatures, for they are pure snow-white, marked with a few blue spots on their summer coats and in winter are covered with long white hair. Their noses are black- - so black that it makes the white look whiter than ever.

Illustrations

The head of one of Sir Claud Alexander's last of the famous Chartley bulls shown below will give a good idea of these beautiful creatures, and the picture of the cow and calf speaks for itself.

Note: White Park cattle are now very scarce.

Head of Horned Park Cattle Bull

Cow and Calf Park Cattle

SHORTHORNS (Horned).

Two kinds: Beef Milk (dual purpose).

Colours: Reds to white and roans.

Descended from the Old Durham cattle, known as Teeswaters, because they were bred in the neighbourhood of the Tees, the old Durhams were gradually improved by the breeders Charles and Robert Colling, also by Bakewell, Bates, Booth and Amos Cruickshank, as well as by many other less known breeders.

Whilst originally a dairy breed, gradually the milking qualities were lost as the beef qualities were developed. The result is an exceptional beef animal, which has become world-famous.

In 1905 the Dairy Shorthorn Association was formed to protect the milking side of the breed; and Shorthorns are therefore divided into the Dairy Shorthorns, which are the dual purpose animals, and the Beef Shorthorns, which are beef only.

BEEF SHORTHORNS

All over the world the Shorthorn is known as Beef, not only when kept as a pure breed, but also as one peculiarly suitable to grade up other cattle. The demand for the best is keen, and high prices have been paid for Shorthorns, both here and abroad.

Apart from high prices at fashionable sales and for export, the average price obtained for stock is very good; and beef Shorthorn breeders are faced with a constant demand.

Whilst Shorthorns are often termed the Red and White breed, the colour varies from pure reds to pure whites, but

Shorthorn Cow (Lady Pamela)

perhaps the favourite colour are the Roans, and good Roans make high price.

The Beef Shorthorn is a broad, well set up animal, able to carry a maximum of flesh on comparatively short legs. Their horns, as their name describes, are short, and stand at right angles to the head, slightly inclined forward and downwards. The nose in the Shorthorn is white and should be free from dark spots.

The approximate weight on Smithfield Club Show analysis showed that Steers under two years weighed 1,415 lb., whilst at three years 1,835, and over three years 2,040 lb.

The Beef Shorthorn matures rapidly, and the usual methods of feeding are employed; in some herds where showing stock is developed, nurse cows are used. A nurse cow is a cow kept for the purpose of feeding young bulls. .

These nurse cows at the Royal Show are always a cause of many enquiries and much speculation.

Development

The Shorthorns were developed by a number of breeders, but usually the major credit goes to Charles and Robert Colling, brothers who farmed separately in the county of Durham. Charles studied Robert Bakewell's methods, as far as was possible, but stumbled upon the value of inbreeding by accident, and went on from there to improve the breed.

Note: This fine breed is quite scarce.

BREEDS OF CATTLE

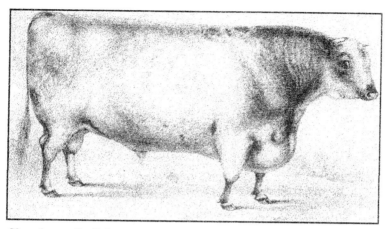

Shorthorn Bull from an old engraving 1846 named *Bellville*
An outstanding bull which took the most important prizes
for four years at the top shows.

Shorthorn Bull and cows 1850

DAIRY SHORTHORNS

The Dairy Shorthorns are good milking cattle, and without considering registered stock, we may say that many of of the ordinary farmers' cows in England are non-pedigree Dairy Shorthorns. They are a popular farmer's breed, because they are dual purpose cattle, fattening when dry, and the bull calves make excellent stores. They are also well built, and have plenty of room to carry flesh, besides being milkers.

It is interesting to know that only a mere coincidence saved England from losing the best of the breed; for it so happened that when Mr. Amos Cruickshank was 83 years old he decided to sell his herd, and to save trouble, sold them all to the Argentine. After some had already left our shores, trouble broke out in South America, and this held up the rest of the herd.

Charles and Robert Colling
Pioneer Breeders

BREEDS OF CATTLE

Dairy Shorthorn   *Dolphin lee Waterloo*
This cow gave 14, 547 lb. in 52 weeks.

SOUTH DEVONS
(Horned : a milk and beef breed)
Colour : Yellow-brown.

These cattle, often termed *South Hams*, are the largest cattle in the world -- huge animals, the bulls weighing as much as 30 cwt. each and the calves looking like grown-ups. Their length is extraordinary; from nose to tail a bull may be nine feet long. They have curly hair. The weight of a cow is 14 to 16 cwt. A great beef breed, and yet milkers, giving high yields of milk with high butterfat percentages.

The history of this breed is unknown, but it is clearly a very old pure variety which stands by itself and has no competitors in its own world.

In colour they are a yellow-red, a curious shade which at once attracts attention, a colour which suggests Guernsey blood in their ancestry. The breed has not developed as well as might have been expected, perhaps mainly because they are too far away from the main parts of our islands.

The ' Ham' breeders claim that these cattle not only mature early, but that their flesh has an extraordinary high proportion of lean to fat, and hence is in special demand where these qualities are recognized.

The demand for export was quite good and it is said that these cattle stand South African conditions exceedingly well and graze successfully on the South African veldt, making the best of the climatic and food conditions. Those interested in live stock should see the South Devons, the Giants of the Cattle world, and compare them with the Dexters, the smallest cattle we have.

62 BREEDS OF CATTLE

South Devon Bull Champion *Bowden Strawberry Boy*

A Beautiful Scene

SUSSEX

(Horned: a beef breed).

Colour: Blood red.

This breed has been established in the south-eastern corner of England from the earliest times, and until within the memory of some people still alive to-day, they were used as draught cattle.

As draughts it is recorded that they had few equals and many an old farmer laments their passing, and asserts that the horse scarcely proves their equal in the clayey lanes of Kent and Sussex, where great strength and steady power is needed.

Except for a few strains, the Sussex has been bred for the single purpose of beef, and as such it has a high reward.

Past recorded figures are:

Under 2 years old   1,430 lb.
"   3   "   1,850 lb.
Over   3   "   2,040 lb.

On several occasions Sussex Steers have exceeded 1 ton in weight. The cattle are of exceedingly fine quality, of a rich blood-red colour, and have a skin resembling the Red Poll in its quality. They have a white bush on the tail, and the horns are long and have black tips.

It is interesting that families in Sussex and Kent have been known to breed the Sussex from generation to generation, and the herd has been handed down with the family heirlooms to be carried on with further enthusiasm and often with great rewards.

64 BREEDS OF CATTLE

Sussex Cattle
Above: Bull: *Imperator*
Below: Cow

WELSH
(Horned: Milk and beef cattle).
Colour :-Black.

When one spoke of Welsh cattle up to quite recently. the allusion would cover a number of types, for each county had cattle peculiar in one or other character.

Many centuries ago references are made to white Welsh cattle with red ears, but this breed has gone.

The Welsh cattle of to-day are black, a few somewhat chocolate coloured, and the blacks may frequently show red-brown patches of rusty-coloured hair in their otherwise black coats. A few years ago the South Wales and North Wales cattle were somewhat different. but to-day very little difference is to be found, except that the Northern cattle have narrower skulls than those of the South. By inter-breeding the various County types have become the present-day Welsh.

You will notice that the cattle are exceedingly well shaped, and have yellow-coloured horns ending in black, which spreads down from the point, whilst some horns are nearly black from head to point.

The horns, apart from their colour, are characteristic for they point forward and up, and are widely set apart; occasionally they have to be trained to take a proper position, for they are liable to get too close together, which spoils the look of the animal.

Although many of the cattle to-day have still the black udders and black teats, a white udder is desired

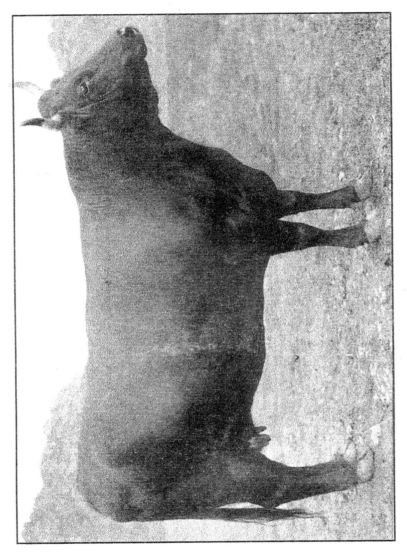

Welsh Cow *Madryn Kate*

BREEDS OF CATTLE

with pink teats; and whilst every effect is made to get this result, great care is taken to prevent any white from spreading in front of the navel.

Welsh calves, which are very pretty; are easily reared, and the cows are good milkers. The Runts, kept as stores in the Midlands, are Welsh blacks, and are very popular, for they thrive on the richer pastures.

White belted Welsh cattle have occasionally been met with, and there is also a White Welsh cattle breed, a very ancient variety, with black noses, ears. eye lashes and feet, and sometimes black spots on the body. Perhaps they are descended from the white cattle with red ears of long ago.

Nancy the Milk Maid

Chaper 2

The
Nature of the Cow

OUR PEDIGREE STOCK.

The home produce question is one that is of great interest to-day, for so much depends on the quantity and quality of our food supply, for it means life, health or disease.

Pedigree Stock-Breeding Centre

Great Britain was essentially the pedigree stock-breeding centre of the world for here by intensive methods, magnificent varieties of stock were produced, so that people came from all over the world to purchase at high prices the best of British cattle.

They are needed not only for pedigree herds, but also to grade up the great herds of cattle kept for beef and cattle in places like the Argentine.

Times change and with difficulties which have faced farmers in recent times, especially diseases, trade has suffered.

HOW CATTLE LIVE*.

The life the cattle lead on the farm and the food they eat is a mystery to many of us, and hardly conceivable to those who for the first time are told of the amount of food consumed by a single animal in a year.

---

* The paragraphs which follow relate to the old style of feeding, what is now loosely called "Organic Feeding", but there have been many developments in animal feeding stuffs and pellets and cake is produced, which makes feeding easier, although in recent times some of the contents have been under scrutiny.

Durham Cattle

These were native to the North of England and were shorthorns. They now appear to be extinct.

One ton of turnips will produce 14 lb. of beef ; a bullock will actually consume 3 cwts. of turnips per day; over a ton a week of turnips alone. In winter over 26 tons of turnips or other roots will be eaten by a single animal on an adequate ration.

Apart from this, cattle need straw and cakes ; cakes made from linseed and from cotton, or from soya beans and other seeds rich in oils.

This is a cow's breakfast and supper in winter time, for a large cow:

Bran 2 lb.	Cotton Cake 3 lb.
Beans 2.50 lb.	Hay 20 lb.
Oats 2.50 lb.	Swedes 20 lb. Cut up.

But everyone has his own method of feeding. It is so often believed that cows on a farm produce milk merely on grass, which people in the towns and cities believe costs nothing. Of course it costs rent, rates, and taxes, and the labour, the cakes and corn, all of which make the production of milk and beef more expensive than people are]ikely to imagine.

The Daily Life

The coming of the calf will start the story. Usually in the early hours of the morning the calf is born, and staggers to its feet, a wet and clumsy little thing, which rolls about with its knees knocking together unable to keep its balance, and tumbles down perhaps.

Hungarian & Scotch Cattle

But young as it is it knows well enough the way to the milk supply, and stands there, shaking and rolling to and fro, nosing round to find the rights and wrongs of it. It soon learns!

The first milk of a cow is called *colostrum,* and is unfit for human food, but it has medical properties which are valuable to the calf. This milk is orange coloured.

Calves soon get their strength and learn how to behave, and a few days later they are taken from the cow, and after that on many farms never see their mother again, as far as the matter of food supply goes.

The farmer or his cowman rears the calf from a bucket and teaches it to drink up the milk, which the calf sucks into its mouth, along a finger inserted in the milk.

It is great fun teaching these "children", for they get so cheeky, so full of life, and they bang their small heads into the bucket, and try to make it give more milk.

On some farms and with some herds, calves are reared on the cows and allowed to have all their mother's milk. They then live the life of the free, running on the pastures by their mother's side.

At two or three months old, the calves have been weaned, except where milk is turned into young beef or where show animals are kept on milk.

Early Hours.

In the early morning in rural England, just as the dawn breaks perhaps, you may meet the cows, waddling home to be milked, walking slowly along the road as if the time was of no consequence.

BREEDS OF CATTLE

Udders
Friesan &
Jersey Cows

They all know their own place in the cowhouse, go to their own spots, except a greedy cow which take a bit from some other cow's heap first before to her own. The rightful owners push this thief and then, amongst it all, the cowman comes, and the delinquent hurries to her stall, with the stolen food dropping from her mouth.

It is a wonderful thing how each cow knows its place in the cowhouse, which holds 40 or 50 or more cows, and where each place looks the same.

They are no sooner settled than the next thing is sound of the milk pouring into buckets, when milking is done by hand. Otherwise, there is the hum of the milking machines. It is the Song of the Cowhouse, with the humming of the men, sound of their voices an accompaniment to the process of the milking. One man can milk eight cows. 10 lb. of milk go to a gallon; about 3 gallons of milk make 1lb. of butter.

Breeds

We have cows which vary in size from the South Devons to the pigmy Dexters. We have cows which give 10 gallons of milk a day: and as a large bucket only holds three gallons, it means three great buckets of milk a day and more.

New Breeds

A number of new breeds have been imported into Britain in recent times, from France and other countries.

Not all agree with these imports and would prefer

Rural Tranquility
Two Cows and a Suckling Calf

BREEDS OF CATTLE

to see the old and traditional breeds being preserved and, where appropriate, improved upon by careful selection.
   The new breeds are:
   1. Charolaise
   2. Main-Anjou
   3. Meuse-Rhine-Issel
   4. Simmental

There are other breeds in various countries, many developed from the traditional British breeds. Whether they will survive, and become old and rare breeds, only time will tell.

INDEX

A
Aberdeen Angus 8
Ayshire 12

B
Bakewell Rbt. 48
Blue Albion 14, 15

C
Calves 73, 75
Colling C & R 59

D
Daily Life 73
Devon (North) 16
Devon (South 61
Dexter 18
Durham Cattle 72

F
Feeding 73
Friesians 73

G
Galloway 26
Guernsey 28
Gloucester 21

H
Hereford 32
Highland Cattle 35

I

J
Jerseys 38

K
Kerry 41

L
Lincoln Reds 43
Longhorns 49

M
Milking 77

N

P
Park Cattle 53
Pedigree Stock 71

S
Shetland 51
Shorthorns 55, 59
Sussex 63

T

U
Udders 76

W
Welsh 65